CONCISE COLLECTION

Modern Commercial Aircraft

Grange BOOKS

Published in 1995
by Grange Books
An imprint of Grange Books Plc.
The Grange
Grange Yard
London SE1 3AG

ISBN 1 85627 762 3

Copyright © 1995 Regency House Publishing Limited

All rights reserved. No part of this publication may be reproduced, stored in a retrieval system, or transmitted in any form or by any means, electronic, mechanical, photocopying, recording or otherwise, without the prior permission of the copyright holder.

Printed in Italy.

Acknowledgements
All photographs from BTPH except:
G. Burridge 27; H. W. Cowin title page, 7, 14, 15, 37.

All artworks supplied by Maltings Partnership

A superb view of a Northwest Airlines Douglas DC 10-40 banking over its home base of Minneapolis/St. Paul.

Contents

Aerospatiale-Aeritalia ATR 42	6
Aerospatiale-Aeritalia ATR 72	7
Airbus Industrie A300	8
Airbus Industrie A310	9
Airbus Industrie A320	10
BAC-Aerospatiale Concorde	11
Boeing 737	12
Boeing 747	13
Boeing 757	14
Boeing 767	15
British Aerospace 125	16
British Aerospace 146	17
British Aerospace ATP	18
British Aerospace Jetstream 31	19
Britten Norman Islander	20
Canadair Challenger 601	21
CASA-IPTN CN-235	22
Cessna Citation II	23
Cessna Titan	24
De Havilland Canada Dash 7	25
De Havilland Canada Dash 8	26
Dornier Do-228	27
Embraer EMB-110 Bandeirante	28
Embraer EMB-120 Brasilia	29
Fokker 50	30
Fokker 100	31
Fokker F-27	32
Gates Learjet	33
Ilyushin Il-62	34
Ilyushin Il-76	35
Ilyushin Il-86	36
Lockheed L-100 Hercules	37
Lockheed L-1011 Tristar	38
McDonnell Douglas DC-9	39
McDonnell Douglas DC-10	40
McDonnell Douglas MD-80	41
Saab 340	42
Shorts 360	43
Tupelov Tu-134	44
Tupelov Tu-154	45
Vickers Viscount	46
Glossary and Abbreviations	

Aerospatiale-Aeritalia ATR 42

Aerospatiale, France and Aeritalia, Italy are equal partners in the manufacture of this regional commercial airliner which first flew in 1984. The French manufacture the wing, engine nacelles, instrumentation and avionics while the Italians manufacture the fuselage, tail fin, undercarriage and flight controls. Final assembly takes place in Toulouse, France. The French airline Air Littoral took the first delivery of this aircraft late in 1985.

The ATR 42 is a relative newcomer to the air transport scene. Below is the companies' demonstrator.

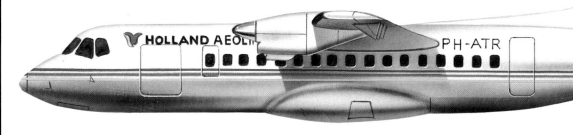

Countries of Origin: France and Italy
Type: medium-range commercial transport
Engines: two 1,800shp Pratt and Whitney Canada PW120 turboprops
Dimensions: span 24.57m (80ft 7in); length 22.67m (74ft 4in); height 7.59m (24ft 11in)
Maximum Take-off Weight: 15,750kg (34,722lb)
Performance: maximum cruising speed 509km/h (316mph); maximum cruising altitude 3,050m (10,000ft); range 1,759km (1,093 miles)
Cockpit Crew: two, plus two cabin
Maximum Seating Capacity: 50

Aerospatiale-Aeritalia ATR 72

The prototype ATR 72 made its maiden flight in October 1988, and the first production example was delivered in 1989. This aircraft is derived from the ATR 42, but its fuselage has been stretched by 4.5m (14ft 9in) which gives space for six extra rows of seats. It also has a greater wingspan which contains more carbon fibre than was used in the ATR 42.

Set to enter airline service in 1989, the photograph below shows a pre-first flight, conceptual model shot cleverly against a city background.

Countries of Origin: France and Italy
Type: medium-range commercial transport
Engines: two 2,400shp Pratt and Whitney Canada PW124 turboprops
Dimensions: span 27.05m (88ft 8in); length 27.17m (89ft 1.5in); height 7.65m (25ft 1in)
Maximum Take-off Weight: 21,500kg (47,400lb)
Performance: maximum cruising speed 530km/h (329mph); maximum cruising altitude 7,620m (25,000ft); range 4,447km (2,763 miles)
Cockpit Crew: two, plus two cabin
Maximum Seating Capacity: 74

Airbus Industrie A300

The A300 entered service with Air France on 23 May 1974, and was the first wide-bodied European airliner to become operational. The A300-600 is one of its later variants and is manufactured by France, Great Britain, Spain and West Germany. It replaced the A300B4, and differs from it by having the new, redesigned fuselage of the A310 which has been extended to seat 18 more passengers. The first JT9D-powered A300-600 flew on 3 July 1983, and Saudia became its initial buyer in May 1984. The first CF6-powered aircraft flew on 20 March 1985, and the first CF6-80C2-powered A300-600 was delivered to Thai International in September 1985.

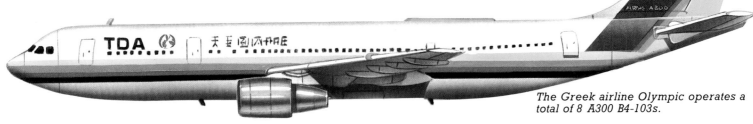

The Greek airline Olympic operates a total of 8 A300 B4-103s.

Countries of Origin: international consortium
Type: medium-range commercial transport
Engines: two 25,400kgp (56,000lb st) Pratt and Whitney JT9D-7R4H1 or General Electric CF6-80C2 turbofans
Dimensions: span 44.8m (147ft 1.25in); length 54.08m (177ft 5in); height 16.53m (54ft 3in)
Maximum Take-off Weight: 165,000kg (363,756lb)
Performance: maximum cruising speed 891km/h (554mph); maximum cruising altitude 10,060m (33,000ft); range 5,200km (3,230 miles)
Cockpit Crew: three, plus six cabin
Maximum Seating Capacity: 344
Notes: details refer to the A300-600

Airbus Industrie A310

In 1978 Airbus Industrie decided to proceed with a variant of the A300 that had a shorter fuselage and a new wing. The first JTD-powered A310-300 flew on 8 July 1985, and was delivered to Swissair that December. The CF6-powered version was delivered to Air India six months later. The A310-300 differs from earlier A310 models in having an additional fuel tank in the tailplane, a carbon-fibre reinforced plastic fin, winglets and a 'glass' cockpit.

Pan Am, in whose livery the Airbus A310-200 below is shown, was the first US customer for the A310 and was the type's only American customer by late 1988.

Countries of Origin: international consortium
Type: medium-range commercial transport
Engines: two 22,680kgp (50,000lb st) Pratt and Whitney JT9D-7R4E or General Electric CF6-80C2-A2, or Pratt and Whitney PW 4152 turbofans
Dimensions: span 43.9m (144ft); length 46.66m (153ft 1in); height 15.81m (51ft 10in)
Maximum Take-off Weight: 150,000kg (330,688lb)
Performance: maximum cruising speed 903km/h (561mph); maximum cruising altitude 11,280m (37,000ft); range 6,950km (4,318 miles)
Cockpit Crew: two or three, plus six cabin
Maximum Seating Capacity: 280
Notes: details refer to the A310-300

Airbus Industrie A320

An advanced passenger jetliner, the A320's order book stood at several hundred prior to its maiden flight. It first flew on 22 February 1987, and Air France took delivery of the earliest production aircraft in June 1988. The first 21 A320s are of the -100 series, but were then succeeded by the -200 series which has an additional fuel tank in the wing's central section as well as winglets. The A320 promises to wrest from the American manufacturers at least some of the market dominance that they have enjoyed since the late 1950s.

One of BA's Airbus A320s, following that airline's acquisition of British Caledonian.

Countries of Origin: international consortium
Type: short/medium-range commercial transport
Engines: two 11,340kgp (25,000lb st) General Electric/SNECMA CFM56-5 or IAE V2500 turbofans
Dimensions: span 33.91m (111ft 3in); length 37.58m (123ft 3in); height 11.77m (38ft 7in)
Maximum Take-off Weight: 71,986kg (158,700lb)
Performance: maximum cruising speed 903km/h (561mph); maximum cruising altitude 11,280m (37,000ft); range 4,730km (2,940 miles)
Cockpit Crew: two, plus four cabin
Maximum Seating Capacity: 164

BAC-Aerospatiale Concorde

Concorde, the world's only operational supersonic airliner, went into service on 21 January 1976 when Air France and British Airways flew to Rio de Janeiro and Bahrain respectively. A total of 18 Concordes were built, with 14 of them entering service. Concorde flies higher and faster than any other commercial aircraft, and its external layout is very unusual because of this. The delta-shaped wings give the aircraft its distinctive shape, and its engines are boxed under the wings in two pairs. Another unusual feature of the aircraft is its 'droop snoot' which can be lowered to give the pilot a better view on take-off and landing, and then raised to provide a steamlined shape for supersonic flight.

The time-shrinking, supersonic, BAC-Aerospatiale Concorde, seen in its original British Airways livery below, cuts the London to New York crossing time to just over three hours, less than half that of a subsonic airliner.

Countries of Origin: Great Britain and France
Type: long-range supersonic commercial transport
Engines: four Rolls-Royce/SNECMA Olympus 593 turbojets
Dimensions: span 25.56m (83ft 10in); length 61.66m (202ft 3.5in); height 11.3m (37ft 1in)
Maximum Take-off Weight: 181,436kg (400,000lb)
Performance: maximum cruising speed 2,179km/h (1,354mph); maximum cruising altitude 18,300m (60,000ft); range 6,300km (3,915 miles)
Cockpit Crew: three, plus six cabin
Maximum Seating Capacity: 144

Boeing 737

The Boeing 737 entered service in early 1968. It competed against the BAC One-Eleven and the DC-9 as a twin-engined, medium-range transport, but unlike them it had underwing engines and an orthodox tail unit. Several variants have been manufactured, and the first 737-400 was rolled out on 26 January 1988 with deliveries to Piedmont taking place in September. The -400 series has more advanced technology than the -300s and its fuselage has been stretched by 2.9m (9ft 6in). The wings and undercarriage have also been strengthened to permit increased landing weights.

By far the best selling of all twin jet airliners: Boeing's short/medium range 737.

Country of Origin: United States of America
Type: short-range commercial transport
Engines: two 9980kgp (22,000lb st) General Electric CFM56-3B-2 or 10,660kgp (23,500lb st) CFM 56-3C turbofans
Dimensions: span 28.9m (94ft 9in); length 36.3m (119ft 1in); height 11.12m (36ft 6in)
Maximum Take-off Weight: 62,824kg (138,500lb)
Performance: maximum cruising speed 912km/h (567mph); maximum cruising altitude 10,670m (35,000ft); range 3,610km (2,244 miles)
Cockpit Crew: two, plus three cabin
Maximum Seating Capacity: 168
Notes: details refer to the 737-400

Boeing 747

The Boeing 747 'Jumbo Jet' is one of the world's most famous aircraft, and is also the heaviest. It proved instantly successful when it entered service with Pan American Airways on 13 December 1969. There have been numerous variants of this wide-bodied jet of which the latest is the 747-400 which made its inaugural flight in March 1988. Northwest Airlines (USA) received the first production aircraft that December. This aircraft differs in several ways from the -300 of which the most noticeable are that the wing has been extended and a vertical winglet has been added to each wing.

Despite Boeing's 747 programme being successful overall, the short-bodied 747SP proved less so.

Country of Origin: United States of America
Type: long-range commercial transport
Engines: four 26,263kgp (57,900lb st) General Electric CF6-80C2, 25,742kgp (56,750lb st) Pratt and Whitney PW 4256 or 26,309kgp (58,000lb st) Rolls-Royce RB 211-524D4D turbofans
Dimensions: span 64.67m (212ft 2in); length 70.67m (231ft 10.25in); height 19.3m (63ft 4in)
Maximum Take-off Weight: 394,630kg (870,000lb)
Performance: maximum cruising speed 939km/h (583mph); maximum cruising altitude 10,670m (35,000ft); range 12,780km (7,940 miles)
Cockpit Crew: two, plus up to 18 cabin
Maximum Seating Capacity: 660
Notes: details refer to the 747-400

Boeing 757

The Boeing 757 is a twin-engined successor to the Boeing 727, and is also a narrow-body airliner that is intended for short and medium ranges. It is, however, much more fuel-efficient. The first one flew on 19 February 1982, and Eastern Airlines (USA) took delivery in December followed by British Airways in January 1983. An extended-range version of the aircraft carries passengers on transatlantic flights, and there are also freighter and 'combi' (which means 'combined passengers and freight') variants.

Illustrated below is one of Northwest Airlines fleet of over 30 Boeing 757s.

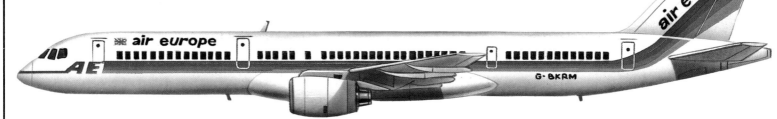

Country of Origin: United States of America
Type: short/medium/long-range commercial transport
Engines: two 17,010kgp (37,500lb st) Rolls-Royce RB 211-535C, 17,327kgp (38,200lb st) Pratt and Whitney 2037 or 18,190kgp (40,100lb st) Rolls-Royce RB 211-535E4 turbofans
Dimensions: span 37.82m (124ft 6in); length 47.47m (155ft 3in); height 13.56m (44ft 6in)
Maximum Take-off Weight: 99,790kg (220,000lb)
Performance: maximum cruising speed 917km/h (570mph); maximum cruising altitude 11,885m (39,000ft); range 8,598km (5,345 miles)
Cockpit Crew: two, plus six cabin
Maximum Seating Capacity: 239

Boeing 767

The Boeing 767 uses a large amount of advanced aluminium lightweight alloys in addition to compounds of graphite epoxy-bonded particles which offer improved anti-corrosion benefits and resist stress better than conventional materials. The programme began in 1978 when 30 of these wide-bodied aircraft were ordered by United Airlines (and these were delivered in 1982). Subsequently, the 767-300 made its inaugural flight in February 1986 and entered service with Japan Air Lines in September. Pilots trained on the Boeing 757 or 767 can fly either aircraft without additional training.

Below is a 767-281 of All Nippon Airways, which operate 30 of the -281s and -381s.

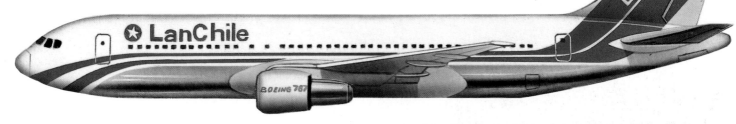

Country of Origin: United States of America
Type: medium/long-range commercial transport
Engines: two 22,680kgp (500,000lb st) Pratt and Whitney JT9D-7R-4E or General Electric CF6-80A2 turbofans
Dimensions: span 47.6m (156ft 1in); length 54.94m (180ft 3in); height 15.85m (52ft)
Maximum Take-off Weight: 159,213kg (351,000lb)
Performance: maximum cruising speed 897km/h (557mph); maximum cruising altitude 11,890m (39,000ft); range 9,305km (5,780 miles)
Cockpit Crew: two or three, plus six cabin
Maximum Seating Capacity: 267
Notes: details refer to the 767-300

British Aerospace 125

The BAe 125 is primarily a corporate transport, but its other roles include acting as a crew trainer, as an aeromedical aircraft and in flight inspection. The prototype made its maiden flight late in 1962, and the aircraft has sold well ever since. The 125-800 entered service in 1984. This has more powerful engines, a new wing-section, new ailerons, a redesigned flight deck and a larger ventral fuel tank.

The latest 125, the -800, which is one of the 710 aircraft sold of all series.

Country of Origin: Great Britain
Type: corporate executive transport
Engines: two 1,950kgp (4,300lb st) Garrett TFE 371-5R-1H turbofans
Dimensions: span 15.66m (51ft 4in); length 15.59m (51ft 2in); height 5.37m (17ft 7in)
Maximum Take-off Weight: 12,430kg (27,400lb)
Performance: maximum cruising speed 845km/h (525mph); maximum cruising altitude 13,100m (42,980ft); range 4,482km (2,765 miles)
Cockpit Crew: two
Maximum Seating Capacity: 14
Notes: details refer to the 125-800

British Aerospace 146

The prototype of the Series 100 first flew in Spring 1982. Five years later, the BAe 146-300 took to the air, and deliveries went to Air UK and Air Wisconsin during the last quarter of 1988, and to Ansett of Australia that December. The aircraft's fuselage is 2.39m (7ft 10in) longer than the -200 series. Both the 146-200 and the 146-300 are available in QT (Quiet Trader) freighter versions, and these have a strengthened floor and freight door. The 146 has been named 'the quietest jet aircraft in the sky'.

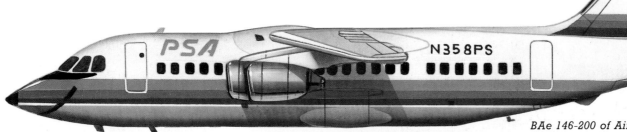

BAe 146-200 of Air Western Australia – an Ansett operating subsidiary.

Country of Origin: Great Britain
Type: medium-range commercial transport
Engines: four 3,161kgp (6,970lb st) Textron Lycoming ALF 502R-5 turbofans
Dimensions: span 26.34m (86ft 5in); length 30.99m (101ft 8in); height 8.61m (28ft 3in)
Maximum Take-off Weight: 43,092kg (95,000lb)
Performance: maximum cruising speed 784km/h (487mph); maximum cruising altitude 9,450m (31,000ft); range 3,467km (2,154 miles)
Cockpit Crew: two, plus four cabin
Maximum Seating Capacity: 110
Notes: details refer to the 146-300

British Aerospace ATP

The first of two prototypes flew on 6 August 1986. Certification was given in January 1988, followed by the aircraft's entry into service with British Midland in June. The ATP (which means 'Advanced Turboprop') is technically a stretched development of the BAe 748. It utilizes new, more powerful engines, systems and equipment, has swept vertical tail surfaces and a redesigned fuselage nose with a 'glass' cockpit.

British Aerospace presently hold firm orders for 26 Advanced Turboprops. Illustrated below is the company's demonstration aircraft.

Country of Origin: Great Britain
Type: medium-range commercial transport
Engines: two 2,400shp Pratt and Whitney Canada PW124A or PW126 turboprops
Dimensions: span 30.63m (100ft 6in); length 26.01m (85ft 4in); height 7.14m (23ft 5in)
Maximum Take-off Weight: 22,453kg (49,500lb)
Performance: maximum cruising speed 492km/h (306mph); maximum cruising altitude 5,485m (18,000ft); range 3,443km (2,140 miles)
Cockpit Crew: two, plus three cabin
Maximum Seating Capacity: 72

British Aerospace Jetstream 31

The first Jetstream 31 flew on 18 March 1982, and deliveries to Contactair began in December. The aircraft is a derivative of the HP 137 Jetstream, which first flew on 18 August 1967. So far, most J31s have been sold in the basic 'commuter' configuration, but an enhanced performance variant designated the Super 31 was introduced in mid-1987: this is fitted with 1,020shp TPE331-12 engines.

The Jetstream 31 has proven extremely popular with North American operators who have accounted for 196 of the 231 aircraft sold, including the one shown below.

Country of Origin: Great Britain
Type: light medium-range airliner and corporate transport
Engines: two 940shp Garret TPE 331-10 turboprops
Dimensions: span 15.85m (52ft); length 14.37m (47ft 2in); height 5.37m (17ft 7in)
Maximum Take-off Weight: 6,950kg (15,322lb)
Performance: maximum cruising speed 482km/h (300mph); maximum cruising altitude 7,620m (25,000ft); range 2,130km (1,324 miles)
Cockpit Crew: two, plus one cabin (optional)
Maximum Seating Capacity: 19

Britten Norman Islander

A small but highly popular transport aircraft, the Islander made its debut in June 1965. There have since been a number of variants of which the most successful has been BN-2S which has a longer nose and a slightly greater passenger capacity. An extension to the basic BN-2A of 1.16m (3ft 9½in) is found on this model, giving it a long, 'probing' nose. By 1974, the Islander had become the best-selling British multi-engined commercial aircraft, and more than 1,100 have now been sold.

Extremely robust and simple to operate, the BN-2 Islander is far and away Britain's best-selling light transport.

Country of Origin: Great Britain
Type: short-haul light transport
Engines: two Avco Lycoming 0-540-E4C5 or 10-540K pistons
Dimensions: span 14.94m (49ft); length 10.86m (37ft 8in); height 4.18m (13ft 9in)
Maximum Take-off Weight: 2,994kg (6,600lb)
Performance: maximum cruising speed 290km/h (180mph); maximum cruising altitude 4,450m (14,600ft); range 1,400km (870 miles)
Cockpit Crew: two
Maximum Seating Capacity: 9
Notes: details refer to the BN-2A

Canadair Challenger 601

The Challenger 601 is a successor to the Challenger 600, and first flew on 10 April 1982. Its most recent variant is the 601-3A which was designed for intercontinental travel. Although most of its initial sales were in North America, this aircraft now sells well internationally.

The second prototype Avco-Lycoming ALF 502-powered Challenger 600 series, photographed below, gained its FAA type certification in November 1980. Deliveries of all versions totalled 162 by March 1988.

Country of Origin: Canada
Type: light corporate transport
Engines: two 4,146kgp (9,140lb st) General Electric CF34-3A turbofans
Dimensions: span 19.61m (64ft 4in); length 20.85m (68ft 5in); height 6.3m (20ft 8in)
Maximum Take-off Weight: 20,230kg (44,600lb)
Performance: maximum cruising speed 815km/h (529mph); maximum cruising altitude 12,500m (41,000ft); range 6,671km (4,145 miles)
Cockpit Crew: two
Maximum Seating Capacity: 19

Casa-Iptn CN-235

The tests for this aircraft took place on two continents, because the first prototype was flown on 11 November 1983 in Spain and the second made its initial flight the following month in Indonesia. A total of 115 aircraft (of both civil and military variants) were on order at the beginning of 1988. Two per month are now being produced (one each from the Indonesian and Spanish assembly lines). The first customer to take delivery was the Indonesian airline Merpati Nusantara in December 1987.

Making a pre-delivery flight, a CN-235 destined for the Caribbean carrier, Prinair.

Countries of Origin: Spain and Indonesia
Type: regional commercial transport and military freighter
Engines: two 1750shp General Electric CT7-9C turboprops
Dimensions: span 25.81m (84ft 7.75in); length 21.35m (70ft 0.5in); height 8.17m (26ft 10in)
Maximum Take-off Weight: 14,400kg (31,745lb)
Performance: maximum cruising speed 452km/h (281mph); maximum cruising altitude 4,570m (15,000ft); range 3,190km (1,994 miles)
Cockpit Crew: two, plus two cabin
Maximum Seating Capacity: 44

Cessna Citation II

This is a twin turbofan executive transport, and the prototype made its first flight in 1969. The Citation was the original production version, eventually to be superseded by the Citation I which had larger wings and more powerful engines. The Citation II also has an increased wingspan and a longer fuselage, and can carry two more passengers.

The photograph shows a characteristic view of a Cessna Citation II operating at about its typical cruising altitude of 10,670m (35,000ft).

Country of Origin: United States of America
Type: short/medium-range business transport
Engines: two Pratt and Whitney Canada JT15D-4 turbofans
Dimensions: span 15.75m (51ft 8in); length 14.38m (47ft 2in); height 4.5m (14ft 9in)
Maximum Take-off Weight: 6,033kg (13,300lb)
Performance: maximum cruising speed 675km/h (420mph); maximum cruising altitude 13,105m (43,000ft); range 3,167km (1,986 miles)
Cockpit Crew: two
Maximum Seating Capacity: 10

Cessna Titan

The Titan can carry a greater payload and has a better performance than the Cessna 402 which it succeeded. Its cabin is convertible, and the aircraft can be transformed into a cargo-carrier, a feeder-liner or a corporate plane. The all-passenger Titan is called the Ambassador, while the utility passenger/cargo variant is known as the Courier. There are, however, no significant external differences between these two. Deliveries of the Titan began in 1976 and continued up until 1982.

The pressurized Cessna Corsair, based on the airframe of the Cessna 421 and now known as the Conquest I.

Country of Origin: United States of America
Type: short-range business transport
Engines: two 375hp Continental TS10-520-J turboprops
Dimensions: span 14.12m (46ft 4in); length 12.04m (39ft 6in); height 4.04m (13ft 3in)
Maximum Take-off Weight: 3,810kg (8,400lb)
Performance: maximum cruising speed 369km/h (229mph); maximum cruising altitude 7,925m (26,000ft); range 2,572km (1,598 miles)
Cockpit Crew: two
Maximum Seating Capacity: 10

De Havilland Canada Dash 7

This aircraft is able to make very short take-offs and landings and so can operate from airfield runways as short as 610m (2,000ft). Although its main function is to carry passengers, it can also carry a mixture of passengers and cargo. The aircraft first flew in 1975 and went into service in 1977. It operates over short routes and, in addition to its excellent airfield performance, is very quiet in operation.

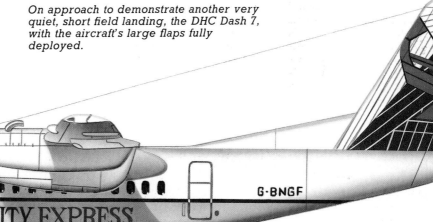

On approach to demonstrate another very quiet, short field landing, the DHC Dash 7, with the aircraft's large flaps fully deployed.

Country of Origin: Canada
Type: medium-range commercial transport
Engines: four Pratt and Whitney Canada PT6A-50 turboprops
Dimensions: span 28.35m (93ft); length 24.58m (80ft 8in); height 7.98m (26ft 2in)
Maximum Take-off Weight: 19,731kg (43,500lb)
Performance: maximum cruising speed 452km/h (280mph); maximum cruising altitude 7,139m (23,600ft); range 1,120km (696 miles)
Cockpit Crew: two, plus two cabin
Maximum Seating Capacity: 50

De Havilland Canada Dash 8

The Dash DHC-8 entered service with NorOntair of Canada in October 1984. The Dash 8-300 prototype flew for the first time on 15 May 1987, and the first delivery was made in late 1988. This model's fuselage has been extended by 3.43m (11ft 2in), and it has a larger wingspan, a strengthened undercarriage, and more powerful engines.

Having made its maiden flight some ten days ahead of schedule, in June 1983, the Dash 8 has gone a long way to redress the relatively poor sales performance of the Dash 7.

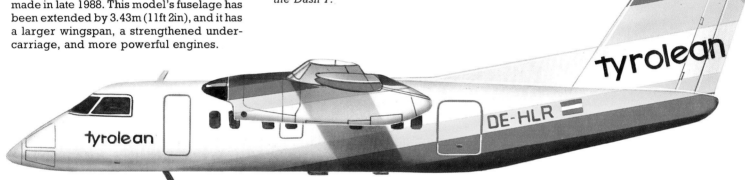

Country of Origin: Canada
Type: medium-range commercial transport
Engines: two 2,830shp Pratt and Whitney Canada PW123 turboprops
Dimensions: span 27.43m (90ft); length 25.68m (84ft 3in); height 7.49m (24ft 7in)
Maximum Take-off Weight: 18,643kg (41,400lb)
Performance: maximum cruising speed 526km/h (327mph); maximum cruising altitude 4,575m (15,000ft); range 1,482km (921 miles)
Cockpit Crew: two, plus two cabin
Maximum Seating Capacity: 56
Notes: details refer to the DHC-8-300

Dornier DO-228

The prototype Do-228-200 flew on 9 May 1981, with A/S Norving of Norway receiving the first production model in August 1982. The aircraft combines a new technology wing of super-critical section with the fuselage cross-section of the Do-128, and two basic versions with a different length of fuselage and range capability are now in production. The 228 is also made under license by Hindustan Aeronautics, India.

Efficient rather than beautiful, the Dornier Do-228 shown here is operated by the Ipswich, Suffolk-based British carrier Suckling Airways.

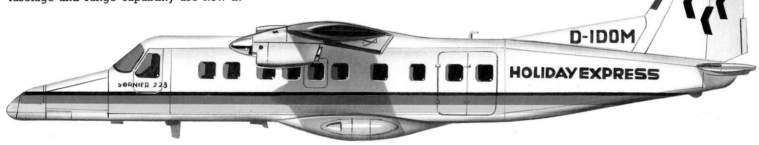

Country of Origin: Federal Republic of Germany
Type: light medium-range airliner and utility transport
Engines: two 715shp Garret AiResearch TPE 331-5-252D turboprops
Dimensions: span 16.97m (55ft 7in); length 16.55m (54ft 3in); height 4.86m (15ft 9in)
Maximum Take-off Weight: 5,700kg (12,570lb)
Performance: maximum cruising speed 432km/h (268mph); maximum cruising altitude 9,020m (29,600ft); range 1,730km (1,075 miles)
Cockpit Crew: two, plus one cabin attendant
Maximum Seating Capacity: 19
Notes: details refer to the 228-200

Embraer EMB-110 Bandeirante

The Bandeirante emerged in 1969 as the first modern light transport of Brazilian origin. It can be seen all over the world flying for commuter airlines, on services to regional locations that are uneconomical for larger aircraft, and replacing such aircraft as the DC-3 in Brazil, where that aircraft can still be seen in service.

An EMB-110P1 Bandeirante of Rio-Sul, a Brazilian-based carrier. Just one of 485 ordered by operators worldwide by the end of 1988.

Country of Origin: Brazil
Type: short-haul corporate transport
Engines: two Pratt and Whitney PT6A-34 turboprops
Dimensions: span 15.32m (50ft 3in); length 15.08m (49ft 6in); height 4.73m (15ft 6in)
Maximum Take-off Weight: 5,670kg (12,500lb)
Performance: maximum cruising speed 422km/h (262mph); maximum cruising altitude 7,350m (24,000ft); range 497km (309 miles)
Cockpit Crew: two
Maximum Seating Capacity: 21

Embraer EMB-120 Brasilia

The first of three prototypes was flown on 27 July 1983, and the first customer delivery to Atlantic Southeast Airlines followed in August 1985. Production was increased from four to five units per month in August 1988. One of the best-selling of the turboprop commuters, the Embraer 120 is also the fastest of the current breed.

Following in the successful footsteps of the Bandeirante, Embraer's 30-passenger EMB-120 Brasilia.

Country of Origin: Brazil
Type: short-haul corporate transport
Engines: two 1,800shp Pratt and Whitney Canada PW118 turboprops
Dimensions: span 19.78m (64ft 10.75in); length 20m (65ft 7in); height 6.35m (20ft 10in)
Maximum Take-off Weight: 11,500kg (25,353lb)
Performance: maximum cruising speed 556km/h (345mph); maximum cruising altitude 7,620m (25,000ft); range 2,982km (1,853 miles)
Cockpit Crew: two
Maximum Seating Capacity: 30

Fokker 50

The Fokker 50 flew for the first time on 28 December 1985, and DLT of West Germany received the first ones on 7 August 1987. The aircraft was based on the Fokker F-27-500 Friendship, and makes much use of composites in its structure. Fokker have also installed new technology engines with an efficient six-blade variant.

More than 110 Fokker 50s had been ordered or optioned by the close of 1988, the one below is in the colours of Ansett, the Australian carrier.

Country of Origin: Netherlands
Type: medium-range commercial airliner
Engines: two 2,250shp Pratt and Whitney Canada PW125B turboprops
Dimensions: span 29m (95ft 1.75in); length 25.19m (83ft 7.75in); height 8.6m (28ft 2.5in)
Maximum Take-off Weight: 20,820kg (45,800lb)
Performance: maximum cruising speed 515km/h (320mph); maximum cruising altitude 7,620m (25,000ft); range 2,938km (1,826 miles)
Cockpit Crew: two, plus two cabin
Maximum Seating Capacity: 60

Fokker 100

The Fokker 100 is a derivative of the F28 Fellowship: it again makes much use of advanced technology, has new systems and equipment, a longer fuselage, and aerodynamically redesigned and strengthened wings as well as new engines from Rolls-Royce. The first of two prototypes made its initial flight on 30 November 1986; the type entering service with Swissair and US Air in 1988.

Aimed in part at replacing the BAC One Eleven, along with earlier DC 9-1 to -30 models, Fokker's twin Rolls-Royce Tay powered 100 was fortunate enough to be selected by Swissair and US Air as co-launch customers, both extremely prestigious airlines.

Country of Origin: Netherlands
Type: short/medium-range commercial transport
Engines: two 6,042kgp (13,320lb st) Rolls-Royce RB183-03 Tay 620-15 turbofans
Dimensions: span 28.08m (92ft 1.5in); length 35.31m (115ft 10in); height 8.6m (27ft 10.5in)
Maximum Take-off Weight: 43,092kg (95,000lb)
Performance: maximum cruising speed 817km/h (508mph); maximum cruising altitude 11,280m (37,000ft); range 4,223km (2,624 miles)
Cockpit Crew: two, plus four cabin
Maximum Seating Capacity: 107

Fokker F-27

Early postwar European airlines approached Fokker with a requirement for a modern airliner in a similar class to the DC-3. The first of two prototypes made its maiden flight on 24 November 1955, and a number of variants have followed all of which have had continually increased fuel capacity and longer range as well as greater fuel efficiency.

One of a total of 785 F27s, including 206 licence-built by Fairchild in the US, were produced. This in the livery of Mississippi Valley Airlines, a carrier that has since been absorbed into Air Wisconsin.

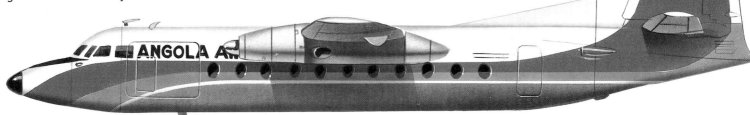

Country of Origin: Netherlands
Type: medium-range commercial airliner
Engines: two 2,105shp Rolls-Royce Dart 528 or 2,230shp Dart 532-7 turboprops
Dimensions: span 29m (95ft 2in); length 25.5m (83ft 7.75in); height 8.41m (27ft 7in)
Maximum Take-off Weight: 19,730kg (43,500lb)
Performance: maximum cruising speed 473km/h (294mph); maximum cruising altitude 8,535m (28,000ft); range 2,660km (1,655 miles)
Cockpit Crew: two, plus two cabin
Maximum Seating Capacity: 48
Notes: details refer to the F27-200

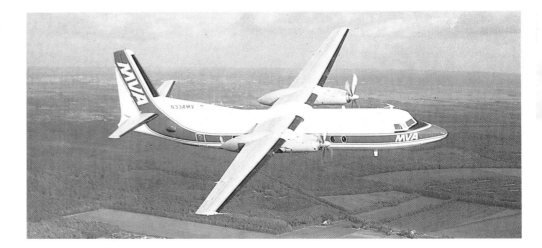

Gates Learjet

The Gates Learjet Longhorn 55 series has a more powerful turbofan engine than its predecessors as well as a cabin that is larger. On 19 April 1979, the first prototype took off from Tucson, Arizona. There are three models of the Longhorn 50 series, all of which have the same basic fuselage, but have increased headroom. Production of the 55 series has now been completed.

Production of the basic model 55, seen below, has now been halted, effort being switched to building the much modified Model 55C.

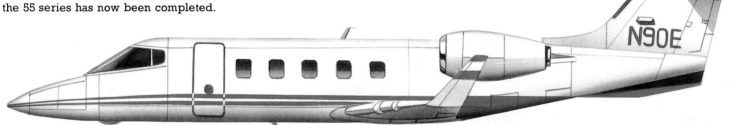

Country of Origin: United States of America
Type: business jet transport
Engines: two Garret TFE731-3-2B turbofans
Dimensions: span 13.35m (43ft 9in); length 14.53m (47ft 8in); height 4.48m (14ft 8in)
Maximum Take-off Weight: 9,290kg (20,500lb)
Performance: maximum cruising speed 743km/h (462mph); maximum cruising altitude 15,545m (51,000ft); range 4,290km (2,666 miles)
Cockpit Crew: two
Maximum Seating Capacity: 10

Ilyushin IL-62

This was designed to complement and eventually replace the Tu-114 on long-distance routes, and first flew in January 1963. It made its public debut at the Paris Air Show in 1965. This was the first long-range four-engined jet developed by the USSR for commercial use. It closely resembled the earlier British-built Vickers VC-10. The aircraft was first used on the Moscow-Montreal flight, and in 1968 began service on the Moscow-New York route which it flies non-stop. The type is mainly in service with the Soviet carrier, Aeroflot, but there are other airlines in Eastern Europe and the Communist countries that also operate it.

The Il-62, still very much the mainstay of Aeroflot's international long-haul services.

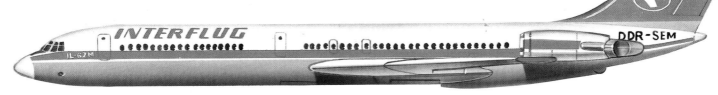

Country of Origin: Soviet Union
Type: long-range commercial airliner
Engines: four 11,500kgp (23,355lb st) Soloviev D-30KU turbofans
Dimensions: span 43.2m (141ft 9in); length 53.12m (174ft 4in); height 12.35m (40ft 6in)
Maximum Take-off Weight: 165,000kg (363,760lb)
Performance: maximum cruising speed 900km/h (560mph); maximum cruising altitude 12,000m (39,000ft); range 10,000km (6,250 miles)
Cockpit Crew: five, plus five cabin
Maximum Seating Capacity: 195

Ilyushin IL-76

Design work on the Il-76 was started in the late 1960s to meet the Soviet Air Force's need for a heavy transport aircraft that would be suitable for operation in the freezing cold of Siberia. The prototype flew for the first time on 25 March 1971, and was displayed at the Paris Air Show two months later. The aircraft is capable of carrying tracked or wheeled vehicles as well as troops. Loading is by means of a ramp at the rear, which has two large clam-shell type doors. The Ilyushin 76 is mainly in service with the Soviet Air Force, although it also flies with Aeroflot, Cubana, Iraqi Airways and Libyan Arab Airlines.

Aeroflot's Il-76s, NATO-codenamed "Candid", are mainly freighters.

Country of Origin: Soviet Union
Type: heavy duty medium/long-range cargo/military aircraft
Engines: four 12,000kgp Soloviev D-30KP turbofans
Dimensions: span 50.5m (165ft 8in); length 46.59m (152ft 10in); height 14.76m (48ft 5in)
Maximum Take-off Weight: 17,000kg (374,790lb)
Performance: maximum cruising speed 850km/h (528mph); maximum cruising altitude 12,000m (39,000ft); range 6,500km (4,040 miles)
Cockpit Crew: seven

Ilyuhin IL-86

The USSR covers a sixth of the globe, and by the 1970s the Soviet Union urgently required a wide-bodied aircraft to fly a large number of passengers over long-distance internal and international routes. Proposals were sought from Antonov, Ilyushin and Tupolev, the contract being awarded to Ilyushin. Subsequently, the first Il-86 flew on 22 December 1976. The aircraft flies on many of Aeroflot's high capacity routes including the very long-range Moscow-Havana service. A unique characteristic is that passengers must enter the aeroplane through doors beneath the cabin deck.

The Il-86 was the first wide-body airliner to be designed and built in the USSR and entered service with Aeroflot in 1980.

Country of Origin: Soviet Union
Type: medium-range commercial airliner
Engines: four 13,000kgp Kuznetsov NK-86 turbofans
Dimensions: span 48.06m (157ft 8in); length 59.54m (195ft 4in); height 15.81m (51ft 10in)
Maximum Take-off Weight: 206,000kg (454,150lb)
Performance: maximum cruising speed 950km/h (590mph); maximum cruising altitude 11,000m (36,090ft); range 4,600km (2,858 miles)
Cockpit Crew: three or four, plus seven cabin
Maximum Seating Capacity: 350

Lockheed L-100 Hercules

The C-130 Hercules was designed by Lockheed to enter service with the United States' Military Air Transport Service (now Military Airlift Command). The prototype made its inaugural flight on 23 August 1954, and production models entered service with the USAF in 1955. Lockheed decided that there was a civilian market for the aircraft, and the L-100-20 was built to fill this need. The L-100-30 is a further stretched version of the -20. More than 1,850 Hercules had been delivered by the end of 1988 to military and commercial users all over the world.

One of the latest L100-30 models, a stretched version of earlier aircraft, of Air Gabon, an African-based operator.

Country of Origin: United States of America
Type: medium/long-range military and commercial freight transport
Engines: four 4,508shp Allison T56-A-15 turboprops
Dimensions: span 40.41m (132ft 7in); length 34.37m (112ft 9in); height 11.66m (38ft 3in)
Maximum Take-off Weight: 70,310kg (155,000lb)
Performance: maximum cruising speed 620km/h (386mph); maximum cruising altitude 6,095m (20,000ft); range 8,617km (5,354 miles)
Cockpit Crew: four

Lockheed L-1011 Tristar

The prototype L-1011 Tristar flew from Palmdale, California on 17 November 1970. Airlines were enthusiastic at first, but this changed once Lockheed and Rolls-Royce (who supplied the engines) ran into financial crises because of this aircraft. Rolls-Royce were bankrupted, and Lockheed nearly so. Only just over 200 aircraft were built, the majority going to Delta and Eastern. However, during the 1980s Cathay Pacific of Hong Kong increased its fleet to become the largest operator outside the USA. Several variants have been built, and Delta Air Lines has re-engined a few Tristars to give greater fuel efficiency.

The pre-delivery flight of a BA Tristar.

Country of Origin: United States of America
Type: long-range commercial airliner
Engines: three 19,050kgp (42,000lb st) Rolls-Royce RB 211-22B or 19,730kgp (43,500lb st) RB 211-22F turbofans
Dimensions: span 47.34m (155ft 4in); length 54.35m (178ft 8in); height 16.78m (55ft 4in)
Maximum Take-off Weight: 195,045kg (240,400lb)
Performance: maximum cruising speed 925km/h (599mph); maximum cruising altitude 12,800m (42,000ft); range 8,080km (5,020 miles)
Cockpit Crew: three, plus eight cabin
Maximum Seating Capacity: 302

McDonnell Douglas DC-9

This aircraft was designed to compete against Sud's Caravelle and the BAC One-Eleven as a short-range turbine-powered aircraft which would have a 'T-tail' and rear-mounted engines. The go-ahead was given for the construction of a prototype without waiting for orders, and it made its maiden flight on 25 February 1965. The aircraft has proved very popular with airlines and passengers alike, and almost a thousand of the type had been sold before production switched to the MD-80. There have been a number of versions of the aircraft.

Operated by Air Holland, this DC 9-32 is capable of operating either as a freighter or passenger liner.

Country of Origin: USA
Type: short/medium-range airliner
Engines: two 6,580kgp (14,510lb st) Pratt and Whitney JT8D-9 or 6,800kgp (14,990lb st) JT8D-11 or 7,030kgp (15,500lb st) JT8D-15 turbofans
Dimensions: span 28.54m (93ft 5in); length 36.37m (119ft 4in); height 8.38m (27ft 6in)
Maximum Take-off Weight: 49,000kg (108,000lb)
Performance: maximum cruising speed 918km/h (572mph); maximum cruising altitude 10,000m (33,000ft); range 1,770km (1,100 miles)
Cockpit Crew: two, plus three cabin
Maximum Seating Capacity: 119
Notes: details refer to the DC-9-30

McDonnell Douglas DC-10

Following the announcement from Boeing that they were proceeding with the 747, it was decided to launch their DC-10 model in direct competition as a transcontinental trijet. Initial design work had begun two years earlier in 1966 when American Airlines had outlined its requirement for a wide-bodied long-range transport. The prototype made its maiden voyage on 29 August 1970, entering service with American Airlines in August 1971. Several variants have entered service.

This DC-10, was leased back from Continental Airlines by NASA and McDonnell Douglas, and was used to test experimental winglets, designed to increase fuel efficiency.

Country of Origin: United States of America
Type: long-range commercial airliner
Engines: three 23,134kgp (51,000lb st) General Electric CF6-50C turbofans
Dimensions: span 50.39m (165ft 4in); length 55.5m (182ft 1in); height 17.7m (58ft 1in)
Maximum Take-off Weight: 259,450kg (572,000lb)
Performance: maximum cruising speed 908km/h (564mph); maximum cruising altitude 10,000m (33,000ft); range 11,580km (7,197 miles)
Cockpit Crew: three, plus eight cabin
Maximum Seating Capacity: 380

McDonnell Douglas MD-80

The DC-9 Series 50 proved popular with airlines, but it was decided that a similar aircraft with a greater capacity and a longer range was required. The prototype was flown on 18 October 1979, and the first production model entered service with Swissair in September 1981. This aircraft has proved very popular, especially with American Airlines, whose single order for one hundred units became the largest requirement for this type from one airline. Also available in this range are the MD-83 and the MD-87.

Illustrated in the photograph below is an MD-87 of Toa Domestic Airlines, now Japan Air System. It was one of 675 MD-80s of all types ordered by the close of 1988.

Country of Origin: United States of America
Type: short/medium-range commercial airliner
Engines: two Pratt and Whitney JT8D-209 turbofans
Dimensions: span 32.87m (107ft 10in); length 45.06m (147ft 10in); height 9.04m (29ft 8in)
Maximum Take-off Weight: 63,502kg (140,000lb)
Performance: maximum cruising speed 898km/h (558mph); maximum cruising altitude 10,670m (35,000ft); range 3,817km (2,372 miles)
Cockpit Crew: two, plus four cabin
Maximum Seating Capacity: 172

SAAB 340

Saab Scania and Fairchild Industries announced in early 1980 their intention to develop a twin-engined turboprop commuter airliner. The prototype flew in 1982 with delivery to the first customer, Crossair, in 1984. Subsequently, Fairchild elected to withdraw from the programme in 1985. The aircraft is proving to be a popular seller, and can be seen throughout the world.

Presented below in close-up is this Northwest Airlink operated SAAB 340 whose affiliates flew 13 out of the 130 machines in service by the close of 1988.

Country of Origin: Sweden
Type: short-range commercial airliner
Engines: two 1,735shp General Electric CT7-5A1 turboprops
Dimensions: span 21.44m (70ft 4in); length 19.72m (64ft 9in); height 6.87m (22ft 6in)
Maximum Take-off Weight: 12,474kg (27,500lb)
Performance: maximum cruising speed 533km/h (288mph); maximum cruising altitude 7,620m (25,000ft); range 3,150km (1,700 miles)
Cockpit Crew: three, plus one cabin
Maximum Seating Capacity: 35

Shorts 360

In 1980, Short Brothers announced its intention to develop a stretched version of its popular SD330, and produced a variant that had an extra six seats as well as a new tail section and improved engines. The aircraft first flew in 1981, and entered service the following year. It is still in production, as is the earlier, shorter and twin-finned 330.

The Shorts 360 has proven very popular with US commuter airlines, this example being destined to operate with American Airlines affiliate, Simmons.

Country of Origin: Great Britain
Type: medium-range commercial airliner
Engines: two 1,424shp Pratt and Whitney Canada PT6A-67R turboprops
Dimensions: span 22.81m (74ft 10in); length 21.59m (70ft 10in); height 7.21m (23ft 8in)
Maximum Take-off Weight: 12,292kg (27,100lb)
Performance: maximum cruising speed 404km/h (251mph); maximum cruising altitude 3,050m (10,000ft); range 1,596km (992 miles)
Cockpit Crew: two, plus one cabin
Maximum Seating Capacity: 39

Tupelov TU-134

Although the Tu-104 and Tu-124 proved popular with Aeroflot, the aircraft did not sell well outside the USSR. It was therefore decided to design a new short-haul medium-capacity airliner for use by Aeroflot and, it was hoped, the overseas market. The Tu-134 entered service with Aeroflot in September 1967, and a lengthened version with upgraded engines that was designated the Tu-134A flew in late 1970. While the Soviet carrier operates the largest number of Tu-134s, a number of Eastern European airlines have purchased the aircraft. A total of 520 Tu-134 airframes were built, and a very large number remain in service.

One of the many Tu-134 jetliners employed by the Soviet Airline Aeroflot on their less densely passenger, intermediate-ranged routes.

Country of Origin: Soviet Union
Type: short/medium-range commercial airliner
Engines: two 6,800kgp (14,990lb st) Soloviev D-30 turbofans
Dimensions: span 29m (95ft 1.75in); length 34.95m (114ft 8in); height 9.02m (29ft 7in)
Maximum Take-off Weight: 45,000kg (99,200lb)
Performance: maximum cruising speed 900km/h (559mph); maximum cruising altitude 12,000m (39,370ft); range 3,500km (2,175 miles)
Cockpit Crew: four, plus four cabin
Maximum Seating Capacity: 80

Tupelov TU-154

In Spring 1966, the Soviet Union released details of a new three-engined aircraft that would replace the Antonov 10, the Ilyushin 18 and the Tupelov 104. It is similar in design to the Boeing 727, but is slightly larger. In October 1968, the Tu-154 made its first flight, and entered service with Aeroflot early in 1971. The aircraft is also operated by other East European carriers.

The trijet Tupolev Tu-154 below belongs to the Soviet airline, Aeroflot. Other operators include Cuba, Czechoslovakia, Poland, Bulgaria and Syria.

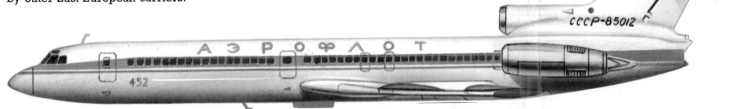

Country of Origin: Soviet Union
Type: medium/long-range commercial airliner
Engines: three 10,500shp Kuznetsov NK-8-2 turbofans
Dimensions: span 37.55m (123ft 3in); length 47.9m (157ft 2in); height 11.4m (37ft 5in)
Maximum Take-off Weight: 96,000kg (211,650lb)
Performance: maximum cruising speed 900km/h (560mph); maximum cruising altitude 11,000m (36,000ft); range 4,000km (2,485 miles)
Cockpit Crew: four, plus four cabin
Maximum Seating Capacity: 164

Vickers Viscount

The most successful post-war British airliner, the Viscount (originally the Viceroy) was conceived near the end of the Second World War, and made its first flight in July 1948. The production model's debut occurred in August 1952. A total of 444 Viscounts were built – the longest production run of a British commercial transport. Several variants were produced which each gave greater passenger loads and more fuel-efficient engines. A few are still in operation, and the largest carrier is British Air Ferries of Southend, Essex.

A Viscount of Lufthansa, which operated the stretched 800 series aircraft.

Country of Origin: Great Britain
Type: short-haul transport
Engines: four Rolls-Royce Dart 525 turboprops
Dimensions: span 28.5m (93ft 8.5in); length 26.1m (85ft 7in); height 8.16m (26ft 9in)
Maximum Take-off Weight: 32,885kg (72,500lb)
Performance: maximum cruising speed 576km/h (358mph); maximum cruising altitude 7,620m (25,000ft); range 2,832km (1,760 miles)
Cockpit Crew: three, plus four cabin
Maximum Seating Capacity: 71
Notes: details refer to the V810